How To
Invent, Protect, & Sell!
The Insider's Guide To Inventing, Product Development, & Marketing

Timothy T. Randolph

BookWorld Press
Sarasota, FL

© 1997 by Timothy Randolph.
All rights reserved.

Published by:
Inventors' Forum Press
P.O. Box 92972
Lakeland, FL 33804
1-800-583-5989

Produced by:
BookWorld Press

Distributed by:
BookWorld Services, Inc.
1933 Whitfield Park Loop
Sarasota, FL 34243
24 hour ordering line: 1-800-444-2524

ISBN: 0-9651527-1-5

Publisher's Cataloging in Publication

Randolph, Timothy.
   How to invent, protect, & sell! : the insider's guide to inventing, product development, & marketing / by Timothy T. Randolph.
   p. cm.
   ISBN 0-9651527-1-5

   1. Inventions—Handbooks, manuals, etc. 2. Patents—Handbooks, manuals, etc. 3. Inventions—Marketing.   I. Title.   II Title: How to invent, protect, and sell!

T339.R36 1996   608
                            QBI96-30056

TABLE OF CONTENTS

INTRODUCTION............................1

PROCEDURES FOR CREATING AN INVENTION.....3
The Literature Review
Patent Protection and Non-Protection
Manufacturing
Finances
Offering Your Invention for Outright Sale
Labels, Packaging and Sales

APPENDICES.............................9
A. Short Stores and Writings from Other Sources.
B. Patent Information.
C. References and Resources.
D. Prototypes.
E. Manufacturers.
F. Invention Awards.
G. Patent Libraries.
H. Inventors' Organizations and Clubs.
I. Sample Letters.
J. Applicable Forms.
K. Partnership Contract.
L. Patent Attorneys.
M. Newspapers and Classified Ads.
N. Mailing Lists.
O. Bulk Rate Mailing.

INTRODUCTION

This book is for the novice, or beginning, inventor. The purpose of the book is to help the beginner avoid undue expenses and frustrations that are often involved in the process of creating an invention or improving upon an existing idea. As an inventor, I have experienced much financial expense and frustration. Because of my experiences as a beginning inventor, I decided to help my fellow beginning inventors avoid many pitfalls that I encountered in the process of creating an invention for market. In this book, I will list and outline the necessary steps that can help you avoid the outlay of large sums of cash and minimize the frustrations that you will encounter in this process.

This valuable reference book is possibly the most complete how-to invention guide. It fills a need of inventors by explaining, step-by-step, licensing, manufacturing, marketing and selling; it also has comprehensive lists of the organizations, such as patent attorneys, patent libraries, prototype makers, etc., necessary to be a successful inventor.

PROCEDURES FOR CREATING AN INVENTION

THE LITERATURE REVIEW

A patent search is necessary to determine if a certain product has been made, or whether you can improve an existing product. Even if you are not a certified agent who can search patents, it is very simple for you to search for products and patents. Later, I will list several patent libraries around the country where you can search at no cost to you. These libraries have a record of all of the patents issued on micro-film, as well as computers and agents to help you get started. These libraries are a direct branch of the patent and trade mark office in Washington, D.C.

If, after you have searched the patent, you do not find the exact product, you can have a patent attorney write a patent for your product. You do not have to have a patent attorney, but if you decide to get one be aware that they can be very costly— as much as $10,000.00. Once you engage a patent attorney, he will research the product; for example, he will speak with you to find out how or why you wanted to invent the product, the kind of material out of which you plan to make this product, and so on. He will write your claims (which will be four [4] or more) concerning the benefits of the product and the functions the product serves.

If you speak or write about your product during the 18 months or so it takes to get a patent from the patent office, you should use the term "patent pending" which means your patent is in the filing stage.

PATENT PROTECTION AND NON-PROTECTION

A patent protects you from anyone stealing your invention and manufacturing the product. When you own the patent, you will be given a patent number. This number should appear on every piece or package you sell or ship. You could manufacture your invention without a patent. To do so, send a drawing and a description of your invention, along with $10.00, to the Patent and Trademark office in Washington, D.C. (The Patent and Trademark Office's address can be found in the back of the book). This will enable the patent office to put your invention on record with you as the first inventor with the idea. In case you make it big and make a lot of money, you can still acquire the patent rights in the United States by doing another search and filing for your patent.

Now you are ready to go and contact manufacturers to make your product. Make sure you send a non-disclosure form to manufacturers before you tell them about your product. Also, before you speak to anybody about your invention, have them sign a non-dis-

Procedures 5

closure form. This will prevent them from talking to anyone else about manufacturing your invention or product for five (5) years.

MANUFACTURING

To find a manufacturer, go to your nearest library. There, you will find a large set of green books entitled: *The Thomas Register*. The set includes twenty-one (21) volumes which list manufacturers throughout the U.S. and Canada. Also, most cities Chambers of Commerce have development councils to steer you to a manufacturer in your county or city. If you contact several manufacturers and are still unable to find a company to take your invention and pay you a royalty, then you should try to negotiate a manufacturer agreement and have it manufactured yourself. Consult the *Thomas Register* again and look for manufacturers to make your invention. In some cases, you may not know the best kind of material to make your product. If your product is being made of plastic, look under "Plastic" and make a list of some companies and their 800 numbers, if they have one. When you call, ask someone in the company if he or she will help you to get it started or steer you in the right direction. They will almost always gladly help you. If the company is near your home, try to make it your business to go and see them. If you have something like a prototype, this will be helpful. In most cases, if this is a new type of product, you will have to buy a mold to make the product. Also, think about what kind of packaging and labeling you are going to use. Later, I will tell you the simple and economic way to sell your product: sell directly to the consumer through ads in newspapers. This way you won't need a marketing person, or fancy packaging.

FINANCES

Inquire with the bookkeeping department of the company you hired to manufacture your product about their credit policies regarding the manufacturing of the product. Try and negotiate an agreement for 30 days net, which means you have 30 days to pay at no interest; or try and negotiate a 2-10 days agreement, which means if you pay in 10 days you receive a 2% discount.

OFFERING YOUR INVENTION FOR OUTRIGHT SALE

If you want to offer your invention to a manufacturer but do not have a patent, you should write the head of the company on your letterhead and send a non-disclosure form to the company; usually, you can obtain the CEO's or president's name from the *Thomas Register*. (See Appendix J for a sample cover letter and non-disclosure form). Often, they will sign and return it, or they may send you their own non-disclo-

Procedures 7

sure form to sign. Sign the non-disclosure form and send it along with the invention you wish for them to review.

If you find a manufacturer with a similar product to your invention, you could ask them to add your product to their line of products. Then, of course, you should ask them for a certain amount of front money. Not every company pays up-front money. In some cases, if they pay front money, your royalties will not exceed 5% on items costing under $10.00. It will be very difficult to get over 5% royalties, but this is good money when you consider all of the manufacturer's overhead. Depending on when they tally-up their sales, the royalties will be paid once per month or once every three (3) months. You would probably engage a company like the Kessler Sales Corporation to work out the terms for you.

LABELS, PACKAGING AND SALES

At a later time, you may want to sell to stores. Then, you will most likely go to a more up-to-date package, with bar-codes and maybe a blister or clam-shell like package. But at first, stay with something simple.

In most cases, when you sell to stores, you will have to adjust your price downward so their mark-up will be reasonable. So at this time, for some immediate cash flow, I would suggest you do your selling by placing a 3 or 4 line classified ad in some of the newspapers listed in the back of this book. If, after spending approximately $150.00 on 3 to 5 ads, you find that it brings in $300.00 in orders, you should then place the same ad in at least 10 papers. Keep tracking your results. You should keep placing the same ad in more newspapers and continue with this ad until you have to hire someone to do this for you. As long as your ad is making money for you, stick with the same ad. If the first ad you placed in the 5 newspapers doesn't generate enough money to double your outlay, rewrite the ad or have a professional ad writer write one for you. (Professional ad writers are listed in the back of the book). I have provided a list of several newspapers from around the country; if you want more newspapers, go to your local library.

APPENDIX A

SHORT STORIES AND OTHER WRITINGS FROM OTHER SOURCES

HOW DO I INVENT?

I try to reduce a problem or situation into a one sentence goal which begins: "How do you?" Once I have the goal on paper, I generally think of all the possibilities. Keeping it simple is the key to inventing. First, I try to find out what is missing, since that's generally where the innovation is. Then I ask whether there is a product vacuum or a niche for an idea. Inventing is moving from the realm of the specialist to that of the generalist.

If you have a new product and are looking for national distribution, you could contact the buyer for one of the major convenience store chains. They may be able to provide a test market for your product and help market it. You could also contact: New Products, 2801 Glenda Ave, Fort Worth, TX 76117. This is a well established mail order company that is looking for new products and who will let you take advantage of their financial strength to sell your product. Contact O.M.I., P.O. Box 9013, Coral Springs, FL 33075.

Lester Nathan, a marketing expert, will provide you with free marketing analysis if you write him at: Results Management and Marketing Inc., 12122 Godfrey Lane, Schenectady NY 12309.

You don't have to be a college graduate to learn the invention business correctly. For example, when Thomas Edison was in the first grade, he was sent home from school after two weeks with a note from his teacher indicating that he was not educable. Yet, this man had over a thousand major patents. If you can think of one good idea, you might be able to come up with 10 or 20. You don't need the thousand Edison had. Read and follow the following instructions and you will become a successful inventor! The world of inventing is vast, there is still plenty left to invent. Begin reading books about inventions, how to market them, how to raise money, etc.

If you want to be a successful innovator, you must start now and learn your trade from professionals. The inventor's society can help you through the process by providing you with educational materials, and sources for product evaluation, funding and marketing. Ask for the "==15 Steps to Successful Inventing==" which has been used to generate major dollars for many inventors. Membership in most inventor clubs or societies is small and they don't ask for up-front fees. The society will steer you in the right direction on how to raise money, how to market your invention, and so forth. Join today!

Appendices

Successful professionals have spent time learning from others. You could do much of the research work yourself; just remember that the society is there for other help and advice.

INVENTORS AND INVENTIONS

Edwin Herbert Land (1909 -) Inventor of the Polaroid Land Camera. In 1927, Harry and Martha Land were delighted with the successes their 18 year-old son had achieved during his first year at Harvard. Imagine their dismay when Edwin announced he wouldn't be returning to Harvard, but would, instead, be heading to New York City to study and experiment. His wished to develop an artificial light polarizer.

Using his college allowance to pay his expenses, Land began experimenting in his tiny New York apartment while he continued his studies at the 2 million volume New York Public Library. What he really needed, however, was a laboratory to conduct more complicated experiments. Being a creative young man, Land, accompanied by a young woman who would later be his wife, discovered a way to make nocturnal visits to a lab at Columbia University. They would climb a fire escape and enter the building through a window that was usually left unlocked. He succeeded in making the first artificial polarizer in that lab, but for the time being, it was useless in any practical sense.

In 1929 Land returned to Harvard, and in a few years, he developed a polarizer formed like an extensive synthetic sheet. He left Harvard in June, 1929, one semester short of graduation, to work in a formal laboratory. His gamble paid off, and on January 31, 1936, the *New York Times* was able to report, "New 'Glass' Cuts Glare of Light. . . . Inventor Shows Polarized Substance." The new substance was called "Polaroid." Polarizing light was valuable, Land explained, because:

> Light from the sun, for instance, is a composite of a great many waves vibrating in all possible planes at right angles to the direction of travel of the beam. Such light is said to be 'unpolarized.' In effect, the waves of the sun are infinitely tangled together. Polarization can be visualized as a complete untangling or 'combing out' of these light waves so that the wave motions all lie in the same plane. 'Polaroid' is the optical comb that accomplishes this effect.

The material, when used in eyeglasses, reduced glare and made it much easier to see. Or, when used with two motion pictures projectors operating in unison, an audience wearing special eyeglasses could experience "3-D" movies. The next year, 1937, Polaroid Corporation was formed.

During World War II, Land and his researchers developed a device for finding the elevation of an airplane above the horizon, goggles that could be used by ski troops, aviation goggles, and, even, goggles for the U.S. fighting dogs who needed protection from flying pebbles. The war effort kept Polaroid's coffers healthy, but Land realized that when the war ended so would the government contracts. He set his scientists on the path of inventing a camera that would not only take pictures but develop them as well. On February 21, 1947, Land astonished members of the Optical Society of America by demonstrating a one-step dry process for producing finished photographs within one minute after taking the picture. These initial photos were sepia toned. During the next several years, the camera was improved and black-and-white instant photographs became a reality. Only fifteen years later, Land and his company debuted Polarcolar Land film, the first instant color film.

From then until now, the history of Polaroid Corporation's growth and success has been built on an evolutionary progression of innovation leading to more automatic and more precisely controlled electronic camera functions. Land eventually acquired 533 patents, second only to Thomas Edison. Although he retired in 1982 as a director and Chairman of the Board of Polaroid, Land has remained, as he always was, the leading force in the company's success. Whatever the challenge, Edwin Land inspired his scientists and engineers with his own sense of determination, tireless work habits, and belief that the answer is there, you just have to find it.

Did you know that a workman who left a soap mixing machine on too long was responsible for making Ivory Soap? He was so embarrassed by his mistake that he threw the mess in a stream. Imagine his dismay when the evidence of his error floated instead of sinking as he had supposed it would. Result: Ivory Soap, the soap that floats.

Did you know that the Band-Aid bandage was invented by a Johnson & Johnson employee whose wife continually cut herself? Earl Dickson's wife was rather accident prone, and he wanted to develop a bandage that she could apply without help. So, he placed a small piece of gauze in the center of a small piece of surgical tape, and what we know today as the Band-Aid was born.

Did you know that the formulas for Coca-Cola and Silly Putty have never been patented? These trade secrets are shared only with trustworthy company employees, and while there have been many attempts to duplicate the products, no one has been successful.

OPPORTUNITY

You've heard the phrase, "Opportunity knocks but once." Don't believe it; opportunity is around every corner, in every nook and cranny, and all you need to do in order to tap into it is to see it. How do you see opportunity? By keeping your mind on the positive, by counting your blessings, by believing in the possibility of all things, and in the possibility of your own success.

When life seems determined to knock you down and

keep you down, don't let it. If you find yourself feeling that things just couldn't get worse, remember that as bad off as you may seem to be you need only to look to your left or to your right and you'll see many others who appear much worse off than you.

A person with a positive attitude sees opportunity everywhere. A person with a negative attitude sees only adversity, and is blind to opportunity. If you really want success you will need not only to commit yourself to that goal, but you'll need to keep a positive frame of mind. ==Don't tell others of your plans for success!== ==Just do it==! When you tell others, you're running the risk of talking your plans into oblivion. Others will be skeptical for a thousand different reasons, sometimes it will be simple jealousy. But whatever the reason, they will almost never be as enthusiastic about your future as you are, and their lack of enthusiasm alone may be enough to kill in you the desire to pursue what may have been a fantastic opportunity. Just do it! After you've made your fortune you'll have plenty of time to talk about it, but you'll probably be too busy having fun.

Success could be attributed to, 'Work like hell!" But you need not kill yourself working. I believe in working smart as opposed to working hard.

INVENTORS' PARANOIAS AND ILLUSIONS

You won't sell your idea by hiding under your bed or giving your life's savings to a lawyer. If there's one thing that all amateur inventors have in common, it's a paranoia: the fear that someone is going to steal their ideas. In fact, it is surprising how many misconceptions new inventors share. These misconceptions work together to keep the typical inventor down, the typical patent attorney's income up, and the typical invention unseen and unsold.

The biggest misconception that amateur inventors have is that their ideas are priceless and so they should patent every one. Inventors often look toward a patent as a goal in itself and think of patents as badges of achievement. Every few months there is an article in the news about some inventor who has hundreds of patents.

Perhaps inventors should start looking at their bank accounts as their measure of achievement. If an idea is priceless, then the inventor should spend money to protect it. But, legal protection has a definable value which is a function of the income an invention can generate.

You should protect patents for inventions that stand to make lots of money. The rule of thumb is this: estimate how many people there are in your target mar-

Appendices

ket. From this you can guess how many people could possibly buy your product.

A good estimate is one out of thirty people. Multiply this number by the wholesale price, and take five percent of that; the invention will make less than this. As you can see, inventions usually make less money than their inventors expect. You need to decide whether this is enough of a return to support the expenditure of possible more than $3000.00 in fees for the utility patent. The best way to reduce the risk is to have lots of ideas. Don't be afraid to take risks and don't worry so much!

MEDICAL PRODUCTS CONSULTANTS

The MPC is a professional group founded by Dr. Arthur Heyman which offers inventors of medical products a means of developing their concept from idea to marketable product. An inventor may require the services of a diverse number of professionals such as lawyers, design engineers, tool and dye makers, as well as needing to be concerned with financial and administrative matters; MPC can coordinate all of these. The organization assists in patent search and application; design and production of original prototypes, artwork, etc.; and subsequent help with promotional activities such as scientific papers and exhibits. MPC is a valuable source for solving product problems for the medical community.

APPENDIX B
PATENT INFORMATION

1. A patent is issued by the patent office for a period of seventeen years.
2. Apply for a patent after you have thoroughly explored all the ways you can apply your idea, which in some cases is a year or two after you thought of the idea. Patenting an idea too soon could be a disaster. You could spend a large amount of money on an idea that you later may discard because you thought of a better way to do it.
3. A "letter patent" or "mechanical patent" is a strong patent. a design that is easy to get around with a minor change will void a design patent or make it unenforceable.
4. It is not necessary to build a working model of your invention, provided you have a good detailed drawing showing how your invention works. You must give a good descriptionof the invention.
5. When you have a production model of your invention it becomes a product.
6. A good, well-kept journal is proof of invention and is a legal document. It can be presented in a court of law to prove you are the inventor. If you keep a sloppy, illegible journal, you will have no backup pro-

Appendices 19

tection for your patent.

COPYRIGHT

Unlike patents and trademarks which are administrated and registered with the U.S. Patent and Trademark Office (PTO), copyrights are registered with the U.S. Copyright Office. For a copy of the brochure "Copyright Basics," write to the Library of Congress, Attention: Copyright Registrar, Washington, DC 20559.

The U.S. Copyright Office has been empowered to issue copyright registrations to authors who have fixed their works into a tangible medium of expression. Examples of works which can be protected by copyright include: literary, musical, pictorial, artistic and motion picture works. The author need do nothing more than fix his work into a tangible medium to create a federal copyright.

To preserve such rights, however, the author must always publish or sell his work with an appropriate copyright notice. Also, unless the author registers his claim to copyright in the copyright office within three months following publication, the copyright owner may lose his remedy for statutory damages and attorney's fees for infringements which took place prior to registration.

The copyright office is not permitted to give legal advice. If you need information or guidance on matters such as disputes over the ownership of a copyright, suits against possible infringes, the procedure for getting a work published, or the method of obtaining royalty payments, you may have to consult an attorney.

The copyright office also examines applications for formal compliance. Before a registration can be issued, the copyright office must satisfy itself that the work exhibits a minimal level of originality. This normally requires a minimal input of creativity. In order to obtain a copyright, an author need not be the "first" to create the work, but must independently create the work without copying from an existing work.

ONE YEAR GRACE PERIOD

An inventor has a grace period of one year (in the U.S.) in which he may place his invention in public use or on sale without losing his right to apply for U.S. patent protection.

COMMONLY ASKED QUESTIONS CONCERNING INTELLECTUAL PROPERTY PROTECTION

1. What is the difference between a design patent and a utility patent?

Appendices

A utility patent covers the concept or idea behind a device or process, whereas a design patent protects only the appearance of the article. The utility patent has a term of 17 years; a design patent is good for 14 years. A design patent application consists primarily of a drawing, whereas a utility patent application includes drawings accompanied by a detailed text and carefully written claims.

2. What are the three requirements for patentability?

The invention must be new, useful and non-obvious. Most patent applications are rejected on the ground that the invention would have been obvious to an imaginative person skilled in that particular art of technology who was aware of all printed material and patents that have ever been published relating to that particular field.

3. If I develop a new idea, must I apply for a patent before I begin selling my product?

No. Although sales or other public disclosures of your invention prior to filing a U.S. patent can cause the loss of foreign patent rights, it will not affect your U.S. patent rights if you file an application within a year of your first offer for sale or other public disclosure.

4. After I apply, how long does it take to get a patent?

Although some patents are issued within a few months, a typical patent takes between 1 and 4 years to be issued, assuming it is ever granted. Some patent applications remain pending for decades.

5. Can I apply for a patent without going through an attorney?

Yes. Several publications exist to assist inventors in filing their own application, including *The Investor's Notebook* by Fred Grissom and David Pressman and *Patent It Yourself* by David Pressman.

6. Is it possible to obtain a patent for an improvement made on a device or process which has already been patented?

Yes. The issues of patentability and infringement are entirely separate. Therefore, one may obtain patent protection for an improvement to a device, yet, to build the improvement and market it in conjunction with the original device, they would infringe the original patent.

7. If I find out that someone is infringing my patent, what will the Patent Office do to protect my rights?

Appendices

Nothing. The Patent Office plays no role in discovering or prosecuting infringes of valid U.S. patents. The patent owner is entirely responsible for bearing the burden and expense of protecting his patent rights.

8.   If two people invent the same thing independently, does the first person to file a patent application receive the patent?

Not necessarily. If two applications are filed claiming the same subject matter, the Patent office begins a special proceeding, known as an "interference," in order to determine who was actually the first inventor. To determine who receives the patent, the Patent Office considers: (i) who was the first to conceive of the invention; (ii) the diligence with which each inventor attempted to reduce his idea to practice; and (iii) who was the first inventor to actually reduce his invention to practice.

9.   If I develop a new product and begin selling it without applying for a patent, can someone else obtain a patent on the idea and prevent me from making my own product?

No. Only the true inventor may apply for a patent.

10.   Can more than one person be named as the inventor in a patent?

Yes. Multiple inventors are quite common, and indeed, it is a legal requirement that all contributors to the inventive concept claimed in the patent be named as inventors.

11. If one of my employees invents something in the course of his duties, can I apply for the patent?

No. Only the true inventor can apply for a patent. However, if the employee develops his invention as part of his job duties, he has a legal duty to assign his entire right in the invention to his employer.

12. If I develop a new, useful and non-obvious method of making something that is already known, can I obtain a patent on the method only?

Yes. Method or process patents are quite common, especially in the fields of chemistry, materials and data processing.

13. If I have a United States patent on some particular apparatus or device, can I prevent someone abroad from making the device and exporting it for sale in the United States?

Your United States patent will not permit you to prevent someone from manufacturing or using your device abroad, but it will prevent the device from being sold or used in the United States, regardless of where it is manufactured.

Appendices 25

14. How can I obtain trademark protection without registering the trademark?

You could either use the trademark in association with the relevant goods or services or take advantage of the trademark "reservation" system.

15. How can I reserve a trademark that I intend to use in the future but have not yet actually begun using?

File a trademark application along with the required "intent to use' statement. Registration of the mark can only take place after actual use of the mark has occurred.

16. If I reserve a corporate name with the Secretary of State, doesn't that give me trademark rights to that name?

No. A corporate name can never take on trademark status until that name is used in association with specific goods and services.

17. If I am using a trademark that is not identical to someone else's trademark, can I still be guilty of trademark infringement?

Yes. Trademark infringement occurs whenever two trademarks exist in the same market which are "confusingly similar" to each other. Thus, if the two trademarks are similar enough to confuse the average person regarding the product's origin or services, trademark infringement has probably occurred.

18. If I obtain a state trademark registration, does that registration guarantee that I have exclusive rights to use the trademark in that state?

No. State trademark laws vary from one state to another, but generally only a cursory examination is performed to determine if your mark is similar to other marks registered in that state. Some states perform no examination whatsoever, and it is quite possible to obtain a state trademark registration for a trademark that is identical to an already existing federally registered trademark. In such a case, a state trademark registration is of little or no value.

19. After I create some literary or artistic work, what do I have to do to obtain a copyright on my creation?

Nothing. Copyright protection comes into being at the moment the work is created. In order to preserve your copyright, you should (but are no longer required to) mark it with a copyright notice, which includes the word "copyright" or "(c)", the year of creation, and your name. Therefore, an appropriate copyright notice could look like: (c) 1986 William Smith.

20. Why would I want to register a copyright if copyright protection comes into being automatically

when I create the work?

Registering a copyright offers procedural advantages if you should ever attempt to prevent the unauthorized copying of your work. Copyright registration may be accomplished by filling out a form available from the Library of Congress and submitting it along with appropriate specimens and fees (usually $10).

21. If I manufacture a product by a secret process and one of my customers discovers that process by analyzing the product, can I recover damages for the theft of my trade secret?

No. A trade secret loses its status as a secret if it can be discovered by members of the public by inspection and analysis of the product. No action may be taken against anyone discovering the trade secret by such methods.

22. What is the address and telephone number of the Copyright Office?

> Copyright Office
> Library of Congress
> 101 Independence Avenue, Room 50
> Washington, DC 20559
> (202) 707-3000

23. What is the address and telephone number of

the U.S. Patent and Trademark Office?

Mailing Address:

U.S. Patent and Trademark Office
Washington, DC 20231

Physical Location:

U.S. Patent and Trademark Office
2011 Jefferson Davis Highway
Crystal Plaza Building #2
Arlington VA 22202
(703) 557-3158

APPENDIX C
REFERENCES AND RESOURCES

THOMAS REGISTER

Thomas Publishing Co.
Thomas Register
1 Penn Plaza N
New York, NY 10017

Most library reference sections contain the 21 volume *Thomas Register*, a reference for information on manu-

Appendices 29

facturers and their products. Directions for using it can be found inside the front and back covers of all volumes.

Volumes 1 through 12 contain an alphabetical listing of various products and services and the names of companies who produce them. Volumes 13 and 14 contain an alphabetical listing of company profiles, along with their addresses, zip codes, telephone numbers, branch offices, asset ratings, and names of company officials. Volume 14 also has a trademark index and an index to the products and services section. Volume 15 through 21, also known as the "Thomcat Catalog File," are an alphabetical listing of catalog information from over 1,250 companies. In addition, the "Thomcat Catalog File" is cross-referenced with volumes 1 through 14. The *Thomas Register* is available for approximately $200.

MANUFACTURER'S REGISTERS

Many library reference sections also carry manufacturer's registers from several states. Most are divided into three major sections: a buyer section which lists products and services alphabetically, an alphabetical section of companies, and a geographical section organized by cities and towns. The geographic section also gives telephone numbers, mailing addresses, and other key information.

Libraries may also have other valuable information

concerning products and manufacturers. You can locate an abundant amount of information by looking in the card file under the general heading "Directories".

If needed, you can obtain a directory of the manufacturing companies in Florida from:

> *Florida Manufacturers Register*
> Manufacturers' News, Inc.
> 4 E Huron Street
> Chicago, IL 60611
> Phone: 312-337-1084

The *Florida Manufacturers Register* is available for approximately $82.

MILLION DOLLAR DIRECTORY

The Million Dollar Directory is a large classified book with about 10,000 pages with every listing you will need. You may call them at: 1-800-526-0652.
Some of their listings include:

— Newspaper publishing or publishing and printing: Page 8432, S.I.C. 2711 (S.I.C. means "Standard Industrial Classification," which refers to a directory published by Dun and Bradstreet).

— Hardware Stores: Page 9324, S.I.C. 5251

Appendices

— Hobby, Toy and Game Shops: Page 9570, S.I.C 5945

— Gift, Novelty and Souvenir Shops: Page 9572, S.I.C. 5947

— Variety Stores: Page 9344, S.I.C. 5331

— Hotels & Motels: Page 990, S. I. C. 7011

— Department Stores: Page 9341, S.I.C. 5311

— Miscellaneous General Merchandise Stores: Page 9347, S.I.C. 5399

— Catalog & Mail Order Houses: Page 9580, S.I.C. 5961

INFORMATION SOURCES

Patent & Trademark Office:
    General Information   703-557-3428
    Public Service Center 703-557-3158
    Commissioner 703-557-3225
    Library 703-557-295
    Copyright Office 202-707-3000
    Library of Congress   202-287-5108
Department of Commerce:
    Library   202-377-3176
    Minority Business Development Agency
————————————————202-377-2000
Department of Energy   202-252-5000

Office of Energy Related Inventions   301-921-3694
National Technical Information Service 703-557-4660
Small Business Administration   800-368-5855
Small Business Innovations Research Programs (SBIR)
———————————————————————202-697-9383
NASA Southern Technology Applications— Center (STAC)
Outside Florida   800-225-0308
Inside Florida 800-354-4832
Government Printing Office   202-783-3238
Consumer Information Center   303-948-3334
Office of Investment, Bureau of Private Enterprise
———————————————————————703-235-1822
Export-Import Bank of U. S   202-566-8320

FLORIDA SMALL BUSINESS DEVELOPMENT CENTERS

Florida Agricultural and Mechanical University
Ms. Patricia McGowan, Director
Small Business Development Center
P.O. Box 708
Tallahassee, FL 32307
Phone: 904-599-3407

Florida Atlantic University
Dr. William Marina, Executive Director
Small Business Development Center
School of Public Administration

Appendices

College of Business & Public Administration
Boca Raton, FL 33431
Phone: 305-338-2273

Fort Lauderdale Small Business Development Center
Mr. William Levi, Associate Director
University Tower
220 Southeast, 2nd Ave
Ft. Lauderdale, FL 33301
Phone: 305-355-5213

Small Business Development Center
Ms. Amber Zentis, Assistant Director
Palm Beach Jr. College
North Campus
3160 PGA Blvd
Palm Beach Gardens, FL 33410
Phone: 305-627-4278

Florida International University
Mr. Marvin Nesbit, Director
Small Business Development Center
Division of Continuing Education
Trailer MO1
Tamiami Campus
Miami, FL 33199
Phone: 305-554-2272

TRADE JOURNALS

Subscribe to trade journals relevant to the focus of your invention. The information, sources, articles and contacts that can be made through your subscription make this a very wise investment indeed.

SERVICES TO HELP THE INVENTOR SUCCEED

Success depends on knowing how to take an invention through the essential steps of protection and completion of prototype, and then to the market place. This is not a simple thing to do. It takes knowledge, determination and perseverance to become a successful inventor.

Inventor's Bookshop (Box 728, Hollywood, FL 33022) has a growing catalog of books and other resource materials about creativity, inventing, self-improvement, business and entrepreneurship. You should start your own personal library of practical reference, how-to and other helpful publications of interest to inventors, innovators, designers, engineers, entrepreneurs and scientists.

DIALOGUE

DIALOGUE is an on-line computer database containing over 200 subjects. Personal computer owners may

Appendices 35

subscribe to this service, but it is fairly expensive. The best way to conduct patent searches using DIALOGUE is to contact a subscriber of DIALOGUE and have them do the search for you. An experienced DIALOGUE searcher can save you money doing the search because you are charged for the time spent on the computer.

The following libraries have the ability to conduct these searches for you. You will need to contact the reference librarian and set up an interview and then arrange a schedule to have the search completed. The search cost will range from $75.00 on up depending upon the computer time used and print charges.

On-line Coordinators at State University Schools (SUS):

Florida Agricultural & Mechanical University
Coleman Memorial Library
PO Box 78-A
Tallahassee, FL 32307
Phone: 904-599-3370   No Contacts Listed

Florida Atlantic University
S.E. Wimberly Library
500 NW 20th St
PO Box 3092
Boca Raton, FL 33431-0992
Phone: 305-393-3760   No Contacts Listed

Florida International University
Bay Vista Campus
Library
North Miami, FL 33181
Phone: 305-940-57301   On-line Services: Susan Weiss

Florida State University
Robert Manning Strozier Library
Tallahassee, FL 32306-2047
Phone: 904-644-5211   No Contacts Listed

University of Central Florida
University Libraries
Orlando, FL 32816-0666
Phone: 305-275-2561
                              On-line Services:
                              Suzanne Holler

BOOKS AND PUBLICATIONS

The following publications include information on how to start a business and on the inventing process:

Artisans & Inventors Report
PO Box 307-B
St Bonifacius, MN 55375
Subscriptions: $12 Annually

Playthings Magazine
51 Madison Ave
New York, NY 10010
(Of interest to toy inventors)

Plastics Technology Magazine
Bill Communications Inc.
633 Third Ave
New York, NY 10017-6743

In Business Magazine
JG Press Inc.
PO Box 323
Emmaus, PA 18049

New Product - New Business Digest
General Electric Co
1 River Rd.
Schenectady, NY 12345

New Products and Processes
Newsweek
444 Madison Ave.
New York, NY 10022

Technology Marketing
General Electric Co
1 River Rd
Schenectady, NY 12345

Agents and Lines Bulletin
5030 Otter Lake Rd.
White Bear, MN 55110
Contains articles and advertisements of manufacturers and their representatives.

INNOVATION CENTERS

There are Innovation Centers all around the country; to get information about them, you need to contact the Small Business Administration. An Innovation Center will do a complete study and give you an evaluation of your invention. They usually give good, sound reports which you should take under consideration when you are finalizing your decision on who to manufacturer your invention.

The Kwikprint Manufacturing Co, Inc.
4868 Victor St
Jacksonville, FL 32207
Hotstamping machines

The Augustine Co Inc
1210 Industrial Blvd
Marshalltown, IA 50158
Dies for stamping machines

Dr. Drorkovitz and Associates
PO Box 1748
Ormand Beach, FL 32175-1748

MARKETING INFORMATION RESOURCES

Jack Hurshman Advertising will produce a very good package or flier for your product for you to send to mail order houses. You may contact him at:

Jack Hurschman
PO Box 649003
San Francisco, CA 94109

Other sources for advertising and promotion pieces:

Jav Reiss Advertising Agency
160-D N Fairview Ave Ste 243
Goleta, CA 93117

New Strategies
Don Lares
3104 E Camelback Rd Ste 526
Phoenix AZ 85016

Kessler Sales Corporation
1247 Napoleon St
Fremont OH 43420

Million Dollar Directory (your local library)

APPENDIX D
PROTOTYPES

ON THE PROTOTYPE

Make your prototype the very best it can be. Ultimately, I had a full size experimental model made. It wasn't as colorful or as sleek and refined as I would have liked, but it worked and was safe. Those two factors are critical. As an inventor, you must weigh each factor in obtaining the best possible prototype. Where children are considered, there is no choice: safety and function are foremost, cosmetic appeal will come later when the product is manufactured commercially.

An additional advantage of having an experimental model is that you can use it consistently and critically, as well as make refinements. This could be critical to your patent search and patent application. Overall, I think that you should be thrifty, creative, experimental and imaginative when making the first prototype of your invention. You can usually "jimmin-rig" with a "make-do" from the hardware store, grocery store, salvation army or your son's junk drawer; but, when it comes time for the experimental model or second prototype, make it as perfect as possible. After all, your invention will have to speak for itself.

For further information on prototypes contact:

Stock Drive Products

55 S Denton Ave.
New Hyde Park, NY 11040

Hexdel Corporation
PO Box 2197
Chatsworth, CA 91311

Economic Development Through Innovation
PO Box 5111
Jackson, MI 39296-5111

MATERIALS AND SUPPLIES

The following catalogs are sources of information to obtain supplies for building prototypes and models:

Aldreich Scientific
PO Box 675
Heltoes, TX 78023
Scientific Supplies
Catalog: $3. 00

Hysol Division
The Dexter Corporation
15051 E Don Juliard Rd.
Industry, CA 91784
Prototyping Supplies

Allied Electronics
401 E 8th St.
Ft. Worth, TX 76102

Electronic components, supplies, tools & equipment
Catalog: $1.00

Manhattan Supply Co.
151 Sunnyside Blvd.
Plainview Long Island, NY 11803
Metal working and precision tools
560 page catalog
Phone: 516-226-4992

Hagenow Laboratories
1302 Washington
Manitowoc, WI 54220
Chemicals and supplies
Catalog: $1.00

Hexcel Corporation - Catalogs
PO Box 2197
20701 Nordhoff St.
Chatsworth, CA 91311
Prototyping Supplies

McKilligan Supply
435 Main St
Johnson City, NY 13790
Tools and supplies for plastic, wood, metal, electronics, graphics and drafting

Stock Drive Products
55 S Denton Ave
New Hyde Park, NY 11040
Prototype and model supplies and parts

Appendices 43

Phone 516-320-0200
900 page catalog: $4.95

Library of Congress
National Referral Center
Washington, DC 20559
The Library of Congress has 13,000 individuals and organizations, who can provide information in various areas, on file.

New Catalog
Department of Commerce
Building 3, Room 1 A02
Washington, DC 20231
Phone: 703-557-2276
You can order the "Guide to the Public Patent Search Facilities of the U.S. Patent and Trademark Office." It tells all about the types and locations of available documents and tells how to do your own patent search.

Internal Revenue Service
Washington DC 20224

"Guide to Free Tax Services" #910
   (Includes an index of free tax publications)
   "Tax Guide for Small Business" #334 "Self Employment Tax" #533
   "Depreciation" #534
   "Business Expenses" #535
   "Tax Information on Partnerships" #541
   "Tax Information on Corporations" #542
   "Business Use of Your Home" #587

"Determining Whether a Worker is an Employee" #SS-8

Food and Drug Administration
5600 Fishers Lane
Rockville, MD 20857
The FDA regulates labeling and packaging of food-related products

APPENDIX E
MANUFACTURERS

The following manufacturers may have an interest in your type of product. Although we try to keep these files current, we know that all files of large size, including this one, will ALWAYS be dated. In the event of a bad address, please write the manufactures number plus the SIC codes on the face of the returned envelope and return it to us. We will purge the name from our list and also pay you $1.00 for each bad address.

HOUSEWARES: HOUSEHOLD & KITCHEN GADGETS, BATHROOM RACKS, CLOSET RACKS WANTED

Ed Teichert
Ekco Housewares
9234 W Belmont Ave
Franklin Park, IL 60131

MFR 9004; SIC 2621

PAPER TOWELS, AND TISSUES MANUFACTURE SEEKS INNOVATIONS, PATENTS.
LOCATION: NORTH AMERICA

Administrator Patent Licensing
Freeman Paper Co Inc.
2954 S Gross
Green Bay, WI 54304

MISCELLANEOUS PLASTICS PRODUCTS . . . FIRM SEEKS LICENSES IN THE PLASTIC PROCESSING FIELD. NOW IN PLASTIC MOLDING, ALSO MAKES TOOTHBRUSHES

Mr. Zieglmayer
M & C Shiffer Co
Adnet
Switzerland
MFR 1500; SIC 3079

PLASTIC PRODUCTS . . . HOUSEHOLD ITEMS. FIRM SEEKS LICENSES NOW PRODUCES BRUSHES, OTHER PLASTIC ARTICLES. MAINLY FOR KITCHEN/BATHROOMS.

Vereinijgte Buerstenfabriken Gmbh
D-78687iodtnau

Schwarzwaldstrasse 15
West Germany

HOUSEHOLD APPLIANCES, FOOD PREPARATION EQUIPMENT, KITCHEN GADGETS. LICENSES SOUGHT FOR MASS-PRODUCTION KITCHEN AND INSTITUTIONAL ITEMS WHICH CAN BE MADE FROM PLASTIC OR METAL. INTEREST INCLUDE ICE CREAM MOLDS, DIPPERS WITH RELEASING DEVICE AND ANY OTHER TYPE OF HIGH-PRODUCTION ITEM IN THIS FIELD. ALSO INTERESTED IN LOW MIXERS & SIMILAR INSTITUTIONAL PREPARATION DEVICES, SMALL HOUSEHOLD HAND & ELECTRIC GARBAGE COMPACTORS

R Steiner
Gitnelli & Co.
Bernstrasse 3052
Zollikofen
Switzerland

MFR 8231; SIC 3079

PRECISION PLASTIC MOLDING . . . LICENSES SOUGHT BY LARGE FIRM

A. A. Finch
Kerr Engineering
Virginia7Street
Greenock PA 15 IFU
Scotland

MFR 8753; SIC 3079

DOOR & WINDOW HARDWARE; LICENSES SOUGHT.

J Green
The Lee Co
1001 S Main
Stillwater, OK 74076
MFR 2139; SIC 3469

METAL STAMPING, NOT ELSEWHERE CLASSIFIED ... HOUSEHOLD ITEMS. MFR OF HOUSEHOLD ARTICLES AND KITCHENWARE IS SEEKING LICENSES FOR THE MFR OF ALL TYPES OF NOVEL KITCHEN & HOUSEHOLD ITEMS. FIRM HAS TWO MODERN FACTORIES AND EMPLOYS 325.

Theodore Klusendick
Koernerstrasse 30
575 Memden, Postrfach 440
West Germany

MFR 8748; SIC 5900

MISC. RETAIL MARKETING GROUP WITH INTERNATIONAL SALES OF OVER 1.5 BILLION SEEKS NEW PRODUCTS IN THE FOLLOWING CATEGORIES: HOME IMPROVEMENT OF ALL TYPES, GARDENING ITEMS, PERSONAL CARE PRODUCTS, RECREATION AND SPORTS PRODUCTS. ALSO INTERESTED IN ELECTRICAL MOTORS AND ELECTRONICS PRODUCTS AND MECHANICAL ITEMS. BUY OR LICENSE PATENTS

M Lucien Berthier
European New Business Officer
BP21
69570 Dardilly
France

MFR 8967; SIC 3088

RACKS, BATHROOM & CLOSET ITEMS SOUGHT

L Remmers
Clairson International
720 SW 17th St
Ocala, FL 35094

MFR 9070; SIC 3089

HOUSEWARES: HOUSEHOLD PRODUCTS

SOUGHT FOR (MARKETING ONLY)

D Noonan
Nationwide Marketing
5628 N Lumberland
Pittsburgh, PA 15217

MFR 9115; SIC 3089

HOUSEWARES: ALL KINDS SOUGHT BY LEADING FIRM LOCATED IN NORTH AMERICA

S. Huff
Rilev M Co Inc
15750 Slkeeler Terrace
Olathe, KS 66062

MFR 9156; SIC 3089

HOUSEWARES; HOUSEHOLD & SPORTING GOOD PRODUCTS SOUGHT

P L Sessions
J H Sessions & Son
PO Box 609E IX Blvd
Bristol, CT 06010

MFR 9063; SIC 3429

HARDWARE: HARDWARE, DRUGSTORE CONSUMER PRODUCTSWANTED

J Gollav
Knickeibocker Corp
2076 N Elston Ave
Chicago, IL 60614

MFR 9063; SIC 3442

HARDWARE: DOOR HARDWARE, NEW ITEMS WANTED FOR MANUFACTURE

G Guirard
Mid-South Door Inc
PO Box 15485
Baton Rouge, LA 70895

MFR 9073; SIC 3442

HOUSEHOLD ARTICLES, PREMIUMS & ADVERTISING SPECIALTIES, PACKAGING ITEMS THAT CAN BE PRODUCED IN PLASTIC. EUROPEAN FIRM SEEKS U.S. PATENTS & PRODUCTS

Plastic A.G.
Bad Ragaz
731.0 Bid Ragaz
Switzerland

MFR 8893; SIC 3079

NOVELTY ITEMS WANTED BY U.S. FIRM

Joel Mallet V/P
Electronic Bonding Inc
200 Jefferson Blvd
Warwick, RI 02888

MFR 8893; SIC 3079

PLASTICS, MOLDED PRODUCTS. U.S. MFR SEEKS EXTREMELY WIDE VARIETY OF PRODUCTS INCLUDING ITEMS MADE OF PLASTIC OR METAL. CAPITAL, MANUFACTURING & MARKETING READILY AVAILABLE

Rick Allen, Pres
BRB Industries
467 Eleventh St
Hoboken, NJ 07030

MFR 8894; SIC 3079

NOVELTIES: SMALL AUSTRALIAN COMPANY SEEKS COMPLETED PRODUCTS FOR DISTRIBUTION BUT ALSO INTERESTED IN PATENT ACQUISITION, IN WHICH CASE THE PRODUCTS WOULD BE CONTRACTED OUT BUT STILL DISTRIBUTED BY HIS FIRM

Robert A Gates
Kros Enterprises Pty Ltd
26 Nurlendi Rd
Melbourne
Australia
MFR 79; SIC 3079

HOUSEWARES: HOUSEHOLD & SPORTING GOOD PRODUCTS SOUGHT

J Pauley
Wistex Inc
2031 Kane St
Lacrosse, WI 546011

MFR 9026; SIC 3429

WINDOW HARDWARE PATENTS WANTED

S Krick
Newill Co
916 S Arcade
Freeport, IL 61032

HOUSEWARES: PROFITABLE NEW ITEMS WANTED FOR MANUFACTURE

G Matthews

American Household Products
1709 Dunnavant Rd SE
Leeds, AL 35094

MFR 9070; SIC 3089

HOUSEWARES: PLASTIC HOUSEWARE, ITEMS SOUGHT BY GROWINGCOMPANY

Administrator Patent Licensing
Tamor Plastics Corp
PO Box 1886, 106 Carter St
Leominster, MA 01453

MISCELLANEOUS COMPANIES

GENERAL HARDWARE MANUFACTURING CO INC
80 White
New York, NY 10013

GENERAL HARDWARE PRODUCTION CO.
75 Van Dyke Ave
Hartford, CT 06106

GRANT HARDWARE CO.
145 High
W Nyack, NY 10994

HAGER HARDWARE CORPORATION
139 Victor St
St Louis, MO 63104

HENNSSGEN HARDWARE CORPORATION
38 Everts Ave Box 2078
Glens Falls NY 12801

IDEAL SECURITY HARDWARE CORPORATION
45 E Maryland Ave
St Paul, MN 55101

INTERNATIONAL HARDWARE INC
10229 Rodney St Box 596
Pineville, NC 28134

JET HARDWARE MANUFACTURING CORPORATION
800 Hinsdale St
Brooklyn, NY 11207

JONES MANUFACTURING CO INC/ HARDWARE DIVISION
Amber Hill Rd Box 1016886247
Birmingham, AL 35210

KEYLIN INC/ HARDWARE DIVISION
13061 E Rosecrans Ave
Santa Fe Springs, CA 90670

KINDER GARD CORPORATION
14822 Ventura Dr

Dallas, TX 75234

HANDY HARDWARE CORPORATION
11 Niagara Ave
Freeport, NY 11520

VSI HARDWARE INDUSTRY
12930 Bradley Ave
Sylmar, CA 91342

WALSALL HARDWARE CORPORATION
7831 E Greenway Rd
Scottsdale AZ 85260

WESSELL HARDWARE CORPORATION
160 S Hartman St
York, PA 17403

WILSON SECURITY HARDWARE
345 Nassau Ave
Brooklyn, NY 11222
LAZY SATHER ROTARY HARDWARE
14189 Meadow Dr
Grass Valley, CA 95945

LEVEY'S HARDWARE
2124 Peach St
Erie, PA 16502
LEVITON MANUFACTURING CO INC
59-25 Little Neck Pkwy
Little Neck, NY 11361

LIBERTY HARDWARE MANUFACTURING CORPORATION
399 Knoolwood Rd Ste 107
White Plains, NY 10603

MAJESTIC LOCK CO INC
194 Daniel
Hackensack, NJ 07601

LITTLESTOWN HARDWARE & FOUNDRY
158 Charles
Littlestown, PA 17340

MARK'S HARDWARE INC
10 Commercial St
Hicksville, NY 11801

CONTINENTAL PLASTIC CO
452 Diens Dr Box 510
Wheeling, IL 60090

PUBLICITY

One of the best publications to publicize your invention in is *Invention Mart*. Thousands of manufacturers from coast to coast read *Invention Mart*. If you would like to have your invention publicized in *Invention Mart*, you can contact them at: Inventors Digest, 2132 E Bijou St., Colorado Springs, CO. 80901

THE JOURNAL

My journal has long been a priceless source of reference to clarify past events, retrieve a name, date, address, phone number or other essential information as well as providing proof of my invention.

HOW TO HANDLE SALES LEADS

Foreign Manufacturers: If you ignore foreign firms, you may be ignoring income. One of our Canadian friends has successfully completed over 72 Licensing Agreements with foreign companies, including some from South Africa. ("Licensing" means that the inventor has granted the manufacturer a "license' to produce his invention, for which the inventor is paid a royalty.

The importance of a foreign patent: Quite a few inventors had marketed overseas without foreign patent protection, and some have had no patents at all. This type of agreement can be done by writing "know-how" contacts, invention-disclosures and non-compete agreements, in which the manufacturer pays an inventor to teach them how to make their product. We spoke with one German manufacturer who said that if the technology looks valuable enough, he would be willing to pay for the airfare for the inventor to present his package, pay for the issuance of the German patent and issue the patent in the name of the inventor, with automatic assignment to the German company upon issuance. This use of automatic assignment is very common in the U.S.A. For example, when an American engineer is being paid to invent something, he is often required to assign the invention to his employer.

Discussion: Deals can be made with overseas companies even if you don't have patent protection. The world is getting smaller, and although it is more difficult than domestic licensing, it's a lot easier than most people think; most of the agreements are even written in English. The advantages of overseas licensing are that you may be able to license your invention in more than one country, thus generating multiple revenues. U.S. Patents are desirable because many firms want to distribute their products in America; but, the laws are now changing and improving to benefit international trade. The disadvantages of overseas licensing are greater difficulty in collecting royalties; inconvenient distances and sometimes lower royalties.

APPENDIX F
INVENTION AWARDS

The B.F. Goodrich Foundation has approved an annual grant to the National Inventors Hall of Fame (NIFH) to create a national invention competition for college students. The program, called the Goodrich Collegiate Inventors Program, began in the fall of 1990.

The foundation has contributed $100,000.00 a year for five years to the NIHF to manage and promote the competition. In addition, the B. F. Goodrich Company has funded up to three awards of $5000.00 each to the best inventions submitted by students. Faculty advisors of winning students were also awarded with grants of $2500.00.

B.F. Goodrich offered to help winning inventors obtain patents at B.F. Goodrich's expense with the understanding that neither NIHF nor B. F. Goodrich would hold any interest in the patents. Winning students and their faculty advisors were invited to Akron, Ohio to receive their awards during the annual induction into the national Inventors Hall of Fame.

Although there are several national contests for primary and secondary grade levels, this was the first one for college students. The B.F. Goodrich Foundation is especially interested in fostering inventive thinking at colleges and universities and in fostering stronger relationships between faculty advisors and students. They know that today's ideas are the seed of tomorrow's success.

INVENTION AWARDS

One Page Entry Format In order to qualify for our Invention Awards, you must submit a one page entry. This must include eight parts.

Part 1 is a photograph or graphic rendering which shows the product in use at the top of a standard 8 1/2 inch by 11 inch page. Parts 2 to 5 are answers to four key question (who, where, what, why) explained below. Part 6 is your name, address, and phone number. Part 7 is your signed declaration that this is your invention and that no confidential information has been revealed. Part 8 is the entry fee which makes the competition self-supporting and free of outside control.

1. The most important part of the entry is the 5 inches (vertical) by 7 inches (horizontal) photograph or graphic rendering of the product in use. There is no substitute for this information. If you can not show the product in use, then it can not be judged — there is no help for it.

If you can not yet show the product in use because that would reveal something you want to keep secret, then you must wait until another competition when you can. Sometimes it is possible to show the product in use but keep the secret part in a "black box." However you do it, you must show the product in use. Be sure to consult your patent attorney before disclosing information.

2. Next you must give a precise, one sentence answer to the question: "Who will buy the product?" To qualify, you must define this market precisely: lifestyle, buying habits, etc.

3. You must also give a precise, one short sentence

answer to: "Where will my customers put it?" Again, the entry will be judged on how precisely you define this marketplace.

4. You must give a precise, one sentence answer to the question: "What problem does the product solve?" The photo or graphic rendering should show most of this. This answer illuminates the key features in the photo or graphic rendering.

5. You must give a precise, one short sentence answer to: "Why will they buy?" Here, you must illuminate the key benefits of your invention, such as reliability, ease of use, cost (which you should have shown in the photo or graphic rendering).

6. Last, you must give your name, address and phone number.

7. Include, date, and sign the following declaration: "I declare that the invention described here is my invention and that I have not revealed any confidential information."

8. Here we had to choose between letting others dictate terms of the competition or asking for a fee. We made the difficult choice of asking you to make the competition self-supporting. Our calculations (guesses?) led us to require a $75 entry fee. If our calculations and guesses are wrong, we will award any surplus to the Top Ten and finalists. Inventors constantly ask us, "I've got an invention. There's nothing

like it. Everybody needs one. What do I do now?" We have been team-teaching this entry format in our workshops for seven years, and it works: Inventors who learn to get this information on one piece of paper do succeed. For further information write to:

Inventor's Digest
2132 E Bijor St.
Colorado Springs, CO 80909-5950

APPENDIX G
PATENT LIBRARIES

Alabama
    Auburn University Libraries
        (205) 826-4500 Ext. 21
    Birmingham Public Library
        (205) 226-3680

Alaska
    Anchorage Municipal Libraries
        (907) 264-4481

Arizona
    Tempe: Arizona State University
        (602)965-7609
    Little Rock: Arkansas Stale Library
        (501) 371-2090

Appendices

California
> Irvine: University of California
> > (714) 856-7234
> Los Angeles Public Library
> > (213) 612-3273
> Sacramento: California State Library
> > (916) 322-4572
> San Diego Public Library
> > (619) 236-5813
> Sunnyvale: Patent Information Clearinghouse
> > (408) 730-7290

Colorado
> Denver Public Library
> > (303) 571-2122

Connecticut
> New Haven: Science Park Library
> > (203)786-5000

Delaware
> Newark: University of Delaware Library
> > (302) 451-2965

Florida
> Fort Lauderdale: Broward County Main Library
> > (305) 357-7444

Miami-Dade Public Library
(305) 375-2665

Georgia

Atlanta: Price Gilbert Memorial Library.
(404) 894-4508

Idaho

Moscow: University of Idaho Library
(208) 885-6235

Illinois

Chicago Public Library
(312) 269-2865
Springfield: Illinois State Library
(217) 782-5430

Indiana

Indianapolis-Marion County Library
(317) 269-1741

Louisiana

Baton Rouge: Louisiana State University
(504) 388-2570

Maryland

College Park: Engineering and Physical Sciences Library, University of Maryland
(301) 454-3037

Massachusetts
    Amherst: Physical Sciences Library, University of Massachusetts
        (413) 545-1370
    Boston Public Library
        (617) 536-5400 Ext. 265

Michigan
    Ann Arbor: Engineering Transportation Library, University of Michigan
        (313) 764-7494
    Detroit Public Library
        (313) 633-1450

Minnesota
    Minneapolis Public Library
        (612) 372-6570

Missouri
    Kansas City: Linda Hall Library
        (816) 363-4600

    St. Louis Public Library
        (314) 241-2288 Ext. 390

Montana
    Butte: Montana College of Mineral Science and Technology Library
        (406) 496-4284

Nebraska

>Lincoln: University of Nebraska-Lincoln, Engineering Library (402) 472-3411

Nevada

>Reno: University of Nevada Library
>(702) 784-6579

New Hampshire

>Durham: University of New Hampshire Library
>(603) 862-1777

New Jersey

>Newark Public Library
>(201) 733-7815

New Mexico

>Albuquerque: University of New Mexico Library
>(505) 277-5441

New York

>Albany: New York State Library
>(518) 474-7040
>Buffalo and Erie County Public Library
>(716) 856-7525 Ext. 267
>New York Public Library
>(212) 714-8529

North Carolina
>Raleigh: O. H. Hill Library,
N.C. State University   (919) 737-3280

Ohio
>Cincinnati & Hamiliton County Public Libraries
>>(513) 369-6936
>Cleveland Public Library
>>(216) 623-2870
>Columbus: Ohio State University Libraries
>>(614) 422-6286

>Toledo/Lucas County Public Library
>>(419) 255-7055 Ext. 212

Oklahoma
>Stillwater: Oklahoma State University Library
>>(405) 624-6546

Oregon
>Salem: Oregon State Library
>>(503) 378-4239

Pennsylvania
>Philadelphia: The Free Library
>>(215) 686-5330
>Pittsburgh: Carnegie Library of Pittsburgh
>>(401) 622-3138

University Park: Pattee Library,
Pennsylvania State University
(814) 865-4861

Rhode Island
Providence Public Library
(401) 521-8726

South Carolina
Charleston: Medical University of South Carolina
(803) 792-2371

Tennessee
Memphis & Shelby County Public Library and Information Center
(901) 725-8876
Nashville: Vanderbilt University Library
(615) 322-2775

Texas
Austin: McKinney Engineering Library, University of Texas
(512) 471-1610

College Station: Sterling C. Evans Library, Texas A & M University
(409) 845-2551

Dallas Public Library
(214) 749-4176
Houston: The Fondren Library,
Rice University
(713) 527-8 101 Ext. 2587

Utah

Salt Lake City: Marriott Library,
University of Utah (801) 581-8394

Virginia

Richmond: Virginia Commonwealth
University Library (804)257-1104

Washington

Seattle: Engineering Library,
University of Washington
(206) 543-0740

Wisconsin

Madison: Kurt F. Wendt Engineering Library,
University of Wisconsin
(608) 262-6845
Miwaukee Public Library
(414) 278-3247

All of the above-listed libraries offer CASSIS (Classification And Search Support Information System), which provides a direct, on-line access to Patent and Trademark Office data.

PATENT DEPOSIT LIBRARIES

The libraries listed below are part of the patent deposit library system. They have in their possession most of the patents that are in the U.S. Patent and Trademark Office and can do on-line searches for patents. If you go to one of these libraries and conduct the search yourself, it will take some time. Plan on spending one to two days in order to locate the right information and do a thorough search.

University of South Florida
4202 Fowler Ave E
Tampa, FL 33620
Phone: 813-974-2011

Broward County Main Library
Government Document Dept.
100 South Andrews Ave.
Ft. Lauderdale, FL 33301
Phone: 954-357-7444

Miami Dade Public Library
Business and Science Department
101 W Flagler St
Miami, FL 13130-2585
Phone: 305-375-2665

Georgia Institute of Technology
Price Albert Memorial Lib
Department of Microforms

Atlanta, GA 30332-0999
Phone: 404-894-4508

APPENDIX H
INVENTORS' ORGANIZATIONS AND CLUBS

Inventors' organizations are the best sources of information for inventors. Local inventor's groups can be found in most large cities. Many of them belong to the National Congress of Inventors Organizations. The main function of local inventor's groups are to help the inventor with product development and business contacts, and to create an informal atmosphere where inventors can meet. A good inventor's group should have a newsletter that announces activities and exhibitions, as well as other news of interest to its members. Inventor's groups are an excellent place to meet your fellow inventors, and to exchange information. They give the inventor an opportunity to show his invention to others and hear the pros and cons of his work.

You may find out about local groups by writing:

The National Congress of Inventor's Organizations
215 Rheem Blvd
Moraga, CA 94556

What the National Congress of Inventor Organizations Does for Inventor Membership

They have one of the larger, if not largest, groups of inventor's organizations in the United States, and we are also a member of the World Intellectual Property Organization (WIPO) and International Federation of Inventor's Association (IFFY) which gives them worldwide distribution. NCI offers a two-way service for the inventors and the Affiliated Business Members by having the following program in effect:

1. NCIO is a non-profit organization which collects the inventions from all the inventor groups who belong to NCIO. They have over two hundred and forty eight inventions worthy of being listed on our mailing list and are continually adding more.

2. As a Business Affiliated Member, your firm can send an outline describing the products, methods, etc., that your company wants developed to NCIO. We send this request for development data to all our invention organizations so that their inventors can solve your companies' development needs. If someone solves your problem, the data is sent to NCIO and they forward the information to your firm. The two of you then get together and negotiate a business deal. If an Affiliated Business Member develops a product that does not fit into its company product line or over-

all company goals, they can still use this service to help them sell their invention.

For non-member groups and/or individuals using our services there is a charge of $200.00 For Inventor Groups, the membership fee is $100.00 per year. For Affiliated Business Members the fee is $600.00 per year. An individual inventor who belongs to a member group is entitled to the services of NCIO in marketing his invention.

INVENTORS' ORGANIZATIONS AND CLUBS

California Inventor's Council
PO Box 2036
Sunnyvale, CA 94087
Barrett Johnson
Phone: 408-732-4314

American Society of Inventors Inc
PO Box 58426
Philadelphia, PA 19102-8426
Harry Sldllman
Phone: 215-546-6601

Inventors of California
National Congress of Inventors Organization
PO Box 158
Rheem Valley, CA 94570

Affiliated Inventors Foundation, Inc
501 Iowa Ave
Colorado Springs, CO 80909
John Faraday
Phone: 303-635-1234

Rocky Mountain Inventors Council
PO Box 5\4365
Denver, CO 80204
Ken Richardson Esq., President

Central Florida Inventors Council
PO Box 13416
Orlando, FL 32859
David E. Flinchbaugh
Phone: 305-859-4835

Inventors of California
National Congress of Inventor's Organizations
PO Box 158
Rheem Valley, CA 94570
Norman Parrish or Gerald McClain
Phone: 415-376-7541

Central Florida Inventors Club
2511 Edgewater Dr
Orlando, FL 32804
Steve Chandler

The Inventors & Entrepreneurs
Society of Indiana, Inc
PO Box 2224

Appendices

Hammond, IN 46323
Prof. Daniel Yovick
Purdue University Calumet
Phone: 219-989-2354

Inventors Assoc. of New England
PO Box 335
Lexington, MA 02173
Don Job, President
Phone: 617-862-5008

Worchester Area Inventors
132 Sterling St
W BoylstoE MA 0 1583
Barbaha Wyatt
Phone: 617-835-6435

Inventors Council of Michigan (INCOM)
2200 Banisteel Blvd
Ann Arbor, MI 48109
J. Downs Herold
Phone: 313-764-5260

American Assoc. of Inventors
6562 E Curtis Rd
Bridgeport, MI 48722
Dennis Martin
Phone: 517-799-8208

Minnesota Inventors Congress
PO Box 71
Redwood Falls, MN 56283

Penny Becker
Phone: 612-253-2344

Society of Minnesota Inventors
PO Box 335
St Cloud, MN 56302
Helen Saatzer
Phone: 612-253-2537

Inventors Assoc. of St. Louis
PO Box 16544
St Louis, MO 63105
Roberta Toole
Phone: 314-534-2677

Albuquerque Invention Club
PO Box 30062
Albuquerque, NM 87190
Dr. Albert Goodman, President
Phone: 505-266-3541

High Technology Entrepreneurs Council (HITEC)
NS Box 72791
Las Vegas, NV 89170
George Sanders
Phone: 702-736-3794

Inventors Club of Greater Cincinnati
18 Gambier Cir
Cincinnati, OH 45218
William M Selenke, Secretary
Phone: 513-825-1222 or 922-9462

Inventors Council of Greater Lorain County
Ohio Technology Transfer Organization
246 Harvard Ave.
Elyria, OH 44035
Dina N Clarke, Sr

Oklahoma Inventor's Congress
PO Box 54625
Oklahoma City, OK 73152
Albert Janco
Phone: 405-848-1991

Invention Development Society
8502A SW 8th St
Oklahoma City, OK 73128
William Enter, Sr
Phone: 215-546-6601

South Dakota Inventors Congress
PO Box 1113
Watertown, SD 57201
Barry Wilfahrt
Phone: 605-886-5814

Tennessee Inventors Assoc.
PO Box 11225
Knoxville, TN 37939
Martin Skinner
Phone: 515-584-0105

Tampa Bay Inventors Council

PO Box 2154
Largo, FL 34294-2254
F. MacNeill MacKay, President
Phone: 813-933-9124 or 813-681-0000

Inventor's Assoc. of Indiana
612 Ironwood Dr
Plainfield, IN 46168
Phone: 317-745-5597

United States Patent Model Foundation
1331 Pennsylvania Ave Ste 903
Washington, DC 20004
Phone: 202-737-1836

Intellectual Property Owners Inc
1255 23rd NW Ste 850
Washington DC 20037
Herberf-Henselv, Exec. Dir.
Phone: 202-466-2396

The Inventor's Club of America Inc
National Headquarters
PO Box 450-291
Atlanta, GA 30345

Inventor's Society of South Florida Inc
PO Box 4306
Boynton Beach, FL 33424

Tampa Bay Inventor Council
805 W 118th Ave

Tampa, FL 33612-4104

APPENDIX I
SAMPLE LETTERS

(Type this letter on your letterhead)

Dear Chief of Research and New Product Development:

I am interested in receiving information regarding your policy for submitting new products for review in order to investigate a potential licensing agreement.

In the event that you do not have an existing standard agreement, I am enclosing a non-disclosure agreement which I would appreciate your signing and returning to me so that I may disclose the product to you for your review.

Thank you very much for your kind consideration of my request. I look forward to a reply at your earliest convenience.

Sincerely,

(Type this letter on your letterhead)

Dear Chief of Research and New Product Development:

I am an entrepreneur with a new product that I believe would fit in your company's product line. I feel this product will greatly increase your revenue. I am interested in receiving information regarding your policy for submitting new products for review in order to investigate a potential licensing agreement.

In the event you do not have an existing standard agreement, I am enclosing a non-disclosure agreement which I would appreciate your signing and returning to me so that I may disclose the product to you for your review.

Thank you very much for your kind consideration of my request. I look forward to a reply at your earliest convenience.

Sincerely,

HOW TO WRITE TO MANUFACTURERS

Appendices 81

1. If you do not know the president's name, address your letter: "Dear President."

2. In the first sentence, tell what your invention does.

3. If you know the market potential of your invention, describe it truthfully.

4. If you know them, describe the production costs.

5. Do not send samples or a copy of your patent in your first letter. In your first letter, you are looking for the right person in the company— not some engineer who is afraid of your invention. A copy of the patent gives such a person the excuse to reject your idea before you have a chance to present it properly.

6. Don't tell them how to make it.

7. Try to get a response from either the president of the company or his direct representative.

8. Enclose a self-addressed, stamped postcard with the message, "Would you like to see more of this interesting product?" ( ) *YES* ( ) *NO*

9. Code the card so you know who responded.

10. Keep trying, over and over, until you get a positive yes or no.

11. NEVER write on lined paper. To make your letter as professional-looking as possible, type it neatly.

12. If you get a "yes" response, contact a person experienced in technology-transfer. Do not use your family lawyer; get an expert. Many patent attorneys are not at all familiar with licensing agreements. Contact either the Licensing Executive Society, which has members in many cities throughout the world, or LIT-CHI, located in Vaduz, Liechtenstein. Virtually all members of both organizations have FAX machines, and can conduct business in English.

APPENDIX J
SAMPLE FORMS AND DOCUMENTS

DISCLOSURE DOCUMENT

On their Disclosure Document Program, I have been informed that when one submits ideas to the Department of Commerce, one obtains a disclosure number which is effective for two years from the date of its reception. Two years pass so quickly; before you realize it, the term of the disclosure will be at hand.

If you have not made your product application in that two year interim, your idea submission is destroyed. If you are the one who has submitted a disclosure document to the patent office in the past twenty two

(22) months, you can simply resubmit a copy of your original disclosure bearing the document number. Send another $10.00 check and request another two year extension. (If you sent a disclosure document more than two years ago, I don't know of a solution).

NON-DISCLOSURE AGREEMENT
COVER LETTER

(Your Letterhead)
Date

Mr. President
(Their address)

Subject: (Your invention name)

Dear Mr.

This letter is pursuant to our telephone conversation on (date) regarding our (your product name) .

This product has been tested by an independent laboratory and the results attest that our product is __% longer lasting (or __% safer, more durable, etc.)

Enclosed you will find two (2) copies of our Non-Disclosure Agreement.
Please sign and return one copy at your earliest convenience.

I will phone you after receiving your signed form; however, if you have any questions, please feel free to call me.

I hope this is the beginning of a long and mutually

profitable association. Respectfully yours,

(Your name)
Enclosures: (2) Non-Disclosure Agreements

NON-DISCLOSURE AGREEMENT

(On Your Letterhead)

Our company agrees that, in consideration for access to information submitted to me or our employees by (your company) , our company will:

1. Keep all information relating to models, drawings, discussions, and printed material in strict confidence within our company.

   2. Disclose this information solely to individuals who have signed a Non-Disclosure Agreement with (your company) or who have written approval from (your company) to receive or have access to the information.

     3.   Not make any contact or agreement of any kind with anyone outside our company on any idea submitted without prior written approval of (your company). Further more we agree neither to use, directly nor indirectly, any such information provided by (your company) for our own benefit or for the benefit of any person, firm or

corporation.

Understood and agreed this _____ day of 19____
Signature (Manufacturer's authorized representative)

Print Name

Title

Company
Signature (Your representative)   Date
Print Name

LICENSE AGREEMENT

A. Parties to This Agreement

THIS AGREEMENT AND LICENSE is made this_____day of_____19_____, by and between:

<u>LICENSER:</u>

having its principal place of business at

and

LICENSEE:

having its principal place of business at

B. Definitions

The following terms, whenever used in this Agreement, shall have the respective meanings set forth below:
(1)   'Licensed Products' shall mean the products of the Licenser set forth herein when made in accordance with licensed know-how or licensed patents as hereinafter defined.
(2)   'Subject Matter of this Agreement shall mean the Licensed Products, any processes for producing the same, or any of them, and any devices for practicing or applying any such processes, or for producing the Licensed Products.
(3)   Licensed know-how' shall mean Licenser's present and future specialized, novel, and unique techniques, inventions, practices, knowledge, skill, experience, and other proprietary information relating to the

Subject Matter of this Agreement.

PATENT COPY REQUEST LETTER

(Your Name)

(Your Address)

Date:
Commissioner of Patents and Trademarks
Washington, D.C. 20231
Subject: Patent Copies
Dear Sir:
Please send me one copy of each of the patents listed below:

Patent #

Inventor

Date

Enclosed is my check # for $ ($1.50 per patent).
Your prompt attention to this request is greatly appreciated.

                          Thank you.
                          Respectfully,

United States Patent [19]
Randolph

[11] Patent Number: 5,008,551
[45] Date of Patent: Apr. 16, 1991

[54] **PHOSPHORESCENT LUMINOUS DOOR KNOBS COVER**

[76] Inventor: Timothy T. Randolph, 1415 Alameda Dr. S., Lakeland, Fla. 33805

[21] Appl. No.: 352,805

[22] Filed: May 16, 1989

[51] Int. Cl.^{5} F21K 2/00
[52] U.S. Cl. 250/462.1; 250/466.1; 16/121
[58] Field of Search 250/462.1, 466.1, 463.1, 250/467.1, 462.1; 542/40; 16/121

[56] **References Cited**

U.S. PATENT DOCUMENTS

1,342,777	6/1920	Thorne	250/466.1
1,522,169	1/1925	Young	250/466.1
1,590,629	6/1926	Julius	16/121
2,032,540	3/1936	Hodny et al.	250/466.1
2,085,331	6/1937	Ramlau	250/466.1
2,155,449	4/1939	Seaman	250/462.1
2,341,583	2/1944	Tuve	250/462.1
2,360,516	10/1944	Schmidling	250/462.1
2,431,169	11/1947	Dice	250/462.1
2,558,433	6/1951	Heinz	250/466.1
2,588,183	3/1952	Vigon	250/462.1
3,174,788	3/1965	Williams	16/121
4,082,351	4/1978	Chrones	16/121
4,745,286	5/1988	Jones	250/462.1
4,832,214	5/1989	Schrader et al.	215/11.1

Primary Examiner—Edward P. Westin
Assistant Examiner—Kim-Kwok Chu
Attorney, Agent, or Firm—Joseph C. Mason, Jr.; Ronald E. Smith

[57] **ABSTRACT**

A luminescent door knob cover. The cover is molded from an admixture of a phosphorescent powder and a carrier. A first half of the cover overlies a first half of a door knob and the first half of a door knob neck when the device is operatively installed. A second half of the cover overlies a second half of the door knob and the second half of the door knob neck when the device is operatively installed. A hinge is formed in the cover to facilitate placing it on and removing it from door knobs. An annular band secures the cover to the door knob.

4 Claims, 3 Drawing Sheets

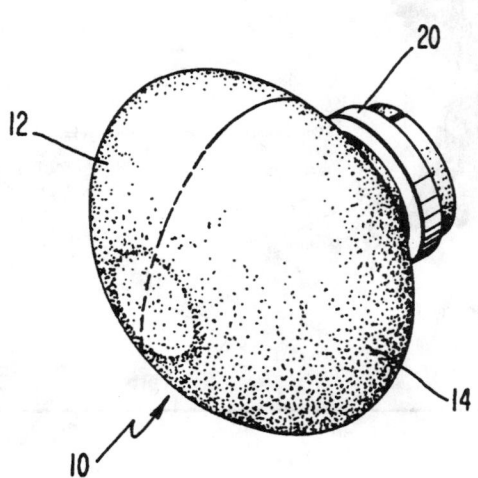

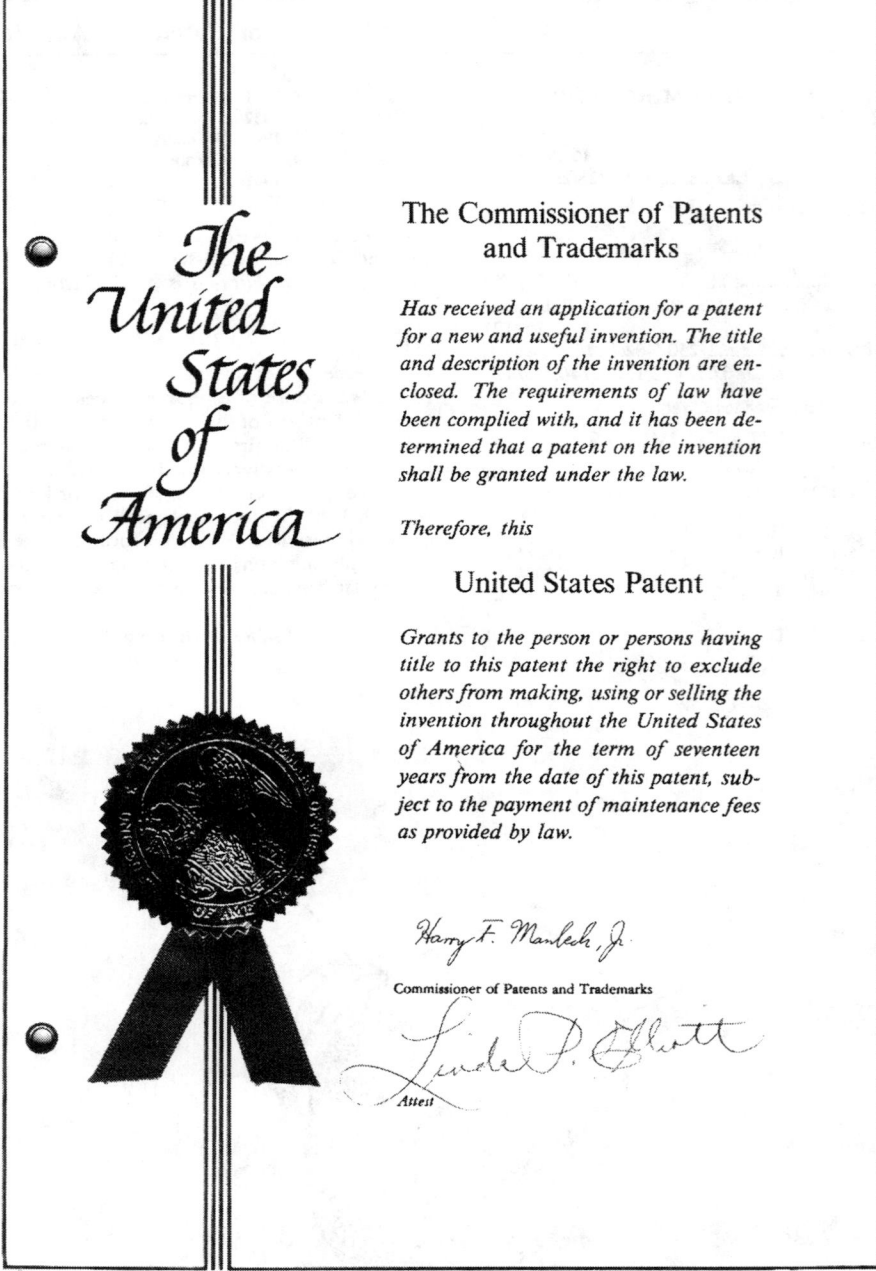

APPENDIX K
PARTNERSHIP CONTRACT

PARTNERSHIP AGREEMENT

This agreement made and entered into this_____day of_____1994

between Timothy Randolph of Lakeland, Florida and _____of _____hereinafter referred to as partners.

WITNESSETH:

WHEREAS: the parties hereto desire to form a Limited Partnership for the purpose of development, manufacture, marketing and sales of illuminated promotional products, namely key chains and door knob covers, and to do such other acts and engage in such other business as the general partner in his sole discretion deems beneficial to the partnership.

NOW, THEREFORE, it is agreed as follows:

1. The parties hereby associate themselves in a Limited Partnership under the name of

2. The principal office of the partnership shall be maintained at Lakeland, Florida or such other location as shall be determined by the general partner.

3. Timothy Randolph shall be the General Partner. A security bond shall not berequired.

4. The General Partner shall devote such time to the partnership as shall be required for its success. No Limited Partner shall participate in the management of the partnership business.

5. All funds of the partnership shall be deposited in such bank accounts or accounts as shall be designated by the General Partner. Withdrawal from such bank account or accounts shall be made upon signature or signatures as the General Partner may designate.

6. The Limited Partner shall contribute five thousand and no/100 ($5,000.00) dollars before or at the execution of this agreement.

7. The Limited Partner shall receive 30% of the net profit of the partnership.

8. The books and records of the partnership shall be available at monthly partnership meetings, the date,

time and place shall be designated by the General Partner. Profits shall be disbursed at said monthly meetings.

9. The partnership shall commence on the date of this agreement and shall continue thereafter for an indefinite period, to expire only by operation of law, by the giving of thirty (30) days written notice by a partner desiring to terminate this agreement or by mutual written consent of the parties.

This agreement shall be construed and enforced in accordance with the laws of the State of Florida, and be binding upon the parties, their executions, successors and assigns.

IN WITNESS WHEREOF, the parties have executed this agreement at Lakeland, Florida, the day and year first above written.

WITNESSES:

(This is only a sample; consult with your attorney.)

APPENDIX L
PATENT ATTORNEYS

Tyner, Earl L
Baldwln & Yeager
1305 Barnett Bank Bldg
904-355-9631

Brooks, Donna
PO Box 11296
Jacksonville, FL 32211
904-721-1986

Kozel, Alexander
5920 San Gabriel Dr
Pensacola, FL 32504
904-477-3461

Corley, Kelly 0
PO Box 273
Gonzalez, FL 32560
904-477-3041

Saliwanchik, David
529 NW 60th St Ste B
Gainesville, FL 32607
904-338-1533

Dominik, Jack Edward
One SE 3rd Ave Ste 2100
Miami, FL 33131
305-358-5667

Kain, Robert C Jr
168 SE First St
Miami, FL 33131
305-374-8418
Martin, Roger L
322 Broadview Ave
Altamonte Springs, FL 32701
305-830-0733

Dumont, Harry R.
415 Montgomiry Rd Ste 175
Altamaonfe Springs, FL 32714
305-788-2789

Norman, Alan H
243 Broadmoor Rd
Lake Mary FL 32746
305-322-9423

Renfro, Julian C
1350 Orange Ave
Winter Park FL 32789
305-628-3600

Allen, Herbert L., Jr.
1 South Orange Ave
Orlando, FL 32801
305-841-2330

Wisner, Carl V Jr
2709 NE 26th Terr
Ft Lauderdale, FL 33306
305-564-2137

Van Der Wall, Robert
2951 S Bayshore Dr Ste 811
Coconut Grove, FL 33133
305-445-6500

Skolnick, Raymond R
10658 180th Ct S
Boca Raton, FL 33434
305-483-6186

McCaghren, Hal H.
120 S Olive Ave Ste 555
West Palm Beach, FL 33458
305-655-1963

Colburn, Harry S H
Alley, Maass, Rogers
321 Royal Poinciana Pl S
Palm Beach, FL 33480
305-659-1770

White, Vincent Augustus
900-5506 US 41N
Brooksville, FL 33512
904-796-5090

Thomas, Bruce K

1420 Northampton Ter
West Palm Beach, FL 33414
305-793-1140

Harris, John L
470 Palm Island NE
Clearwater, FL 33515
813-446-1953

Alvin S Blum Sc. B.
Complete patent application service
Search, drawing, file, prosecute.
2350 Del Mar Place
Ft Lauderdale, FL 33201

David W Pettis Jr
Duckworth, Allen, Dyert, Pettis
PO Box 1528
Tampa, FL 33601
813- 229-8176

Mason, Mason & Associates
Arbor Shoreline Center
1307 US 19 S Ste 102
Clearwater, FL 34624

APPENDIX M
NEWSPAPERS AND CLASSIFIED ADS

The Daily News
220 E 42nd St
New York, NY 10001

Pennysaver
3756 Gross Circle
Carson City, NV 89702

Alaska Mazazine
Classified Advertising
808 E Street Ste 200-
Anchorage, AK 99501

M K Press
PO Box 355
Randolph, MA 02368

National Mail Order Classified
PO Box 5
Sarasota, FL 34230
813-366-3003

Business Opportunities Journal
Business Ventures
1050 Rosecrans
PO Box 60762
San Diego, CA 92166

APPENDIX N
MAILING LISTS

You can order mailing lists for almost all categories. However, some companies that sell mailing lists may have more current names than others. A good way to gauge a mailing list is to order the minimum amount of names from several companies. Keep a written record of what you send to each list. Also, match all orders and returned mail to each list so that you can determine which lists are the most productive. There may be other ways to accomplish this, but I like this way better.

Dale Advertising Co
PO Box 279
Gvover, NC 28073

Landers Herald
720 Morrow Ave
Clayton, NJ 08312

George E. Norr
PO Box 70235
Hunter, UT 84170

Richard Lind
PO Box 6965
Ft. Myers, FL 33991

Jack Hurshman
PO Box 649003
San Francisco, CA 94109

APPENDIX O
BULK RATE MAILING

Go to your main post office in your area and inquire about the "Bulk Rate" mailing procedure. Private bulk rate mailing companies are usually less expensive, generally about 1/2 the normal rate, provided you mail the same merchandise or information to at least 200 people. There will be additional charges for sorting your names by state and zip code. For more detailed information on this subject contact your main post office, or check your local telephone directory under "Direct Mail."

DIRECTORY OF LARGER MAIL ORDER HOUSES

Deerskin Trading Post
119 Foster St.
Peabody, MA 01960

Yield House
PO Box 1000
North Conway, NH 03860

Appendices 101

Popular Services, Inc.
128 Dayton Ave
Passaic, NJ 07055

American Image Industries
276 Park Ave
New York, NY 10010

The Ferry House, Inc.
554 North State Rd
Briarcliff Manor, NY 10510

Giftware Corner
482 Sunrise Why
Rockville Ctr, NY 11571

Rombins' Host Farm
Rt 116 West
Fairfield, PA 17320

Postamatic Company
Lafayette Hill, Pennsylvania 19444

Charles Keith. Ltd
4030 Pleasantdale Rd
Atlanta, GA 30340

Chris & Craft
Algonac, Michigan 48001

Suburbia, Inc.
366 Wacouta
St Paul, MN 55101

House of Minnell
Dearpath Road
Batavia, IL 60510

National Reporter Publ
15115 S 76 E Ave
Bixby, OK 74008

Krupp Mail Order
PO Box 9090
Boulder, CO 80301

Ambassador
711 W Broadway
Tempe, AZ 85282

United States Sales Corporation
9351 Laurel Canyon Blvd.
Pacoima, CA 91331

Wrightwell Co.
10 B Massachusetts Ave
Boston, MA 02115

L L Bean Inc.
481 Main St
Freeport, ME 04032

Appendices

Pat Harris, Inc.
725 Dell Road
Carlstadt, NJ 07072

Nora Nelson, Inc.
621 Ave of the Americas
Now York, NY 10011

Foster-Trent, Inc.
2345 Boston Post Rd
Larchmont, NV 10538

Colonial Garden Kitchens
270 W Merrick Rd
Valley Stream, NY 11562

Hanover House, Inc.
Hanover, Pennsylvania 17331

The Game Room
PO Box 4290
Washington, DC 20012

Joan Cook
851 Eller Drive
Ft Lauderdale, FL 33316

Johnston Smith Co
35075 Automation Dr.
Mt. Clemens, MI 48043

Fingerhut Corporation
4400 Baker Rd
Minnetonka, MN 55343

American Products
5550 N Elston Ave
Chicago, IL 60630

P & S Sales
3818 S 79th East Ave
Tulsa, OK 74145

Bruce Bolind, Inc.
Bolind Bldg.
Boulder, CO 80302

Old Pueblo Traders
PO Box 27800
Tucson, AZ 86726

SM Mail Order Mktg
PO Box 8416
Universal City, CA 91608

Constance Carol
PO Box 899
Plymouth, MA 02360

Shopping International
Norwich, Vermont 05055

Mail Order Associates
120 Chestnut Ridge Road
Montvale, NJ 07645

Charter Guild
641 Lexington Ave
New York, NY 10021

Lillian Vernon
510 S Fulton Ave
Mt Vernon, NY 10550

Sunset House
840 S Broadway
Hicksville, NY 11802

Harriet Carter Gifts
Stump Road
Montgomeryville, PA 18936

Sleepy Hollow Gifts
6651 Arlington Blvd
Falls Church, VA 22042

Nancy Stone
5729 Pearl Road
Cleveland, OH 44129

Down's & Company
6048 W Beloit Rd
Milwaukee, WI 53219

Niresk Industries, Inc
666 Dundee Rd. 0701
Northbrook, IL 60062

Foster & Gallagher.
6523 N Galena Rd
Peoria, IL 61601

Norchow Collection
PO Box 34257
Dallas, TX 75234

Walter Drake & Sons, Inc
Drake Bldg
Colorado Springs, CO 80940

Gift Guido
5818 Venice Blvd
Los Angeles, CA 90019

StarCrest of California
3159 Redhill Ave
Costa Mesa, CA 92626

Clymer's of Bucks County
Chestnut St
Nashua, NH 03060

Grolier Enterprises. Inc.
Sherman Turnpike
Danbury, CT 06816

Contemporary Marketing
790 Maple Lane
Bensenville, IL 60106

Carol Wright Gifts
809 "P" St
Lincon, NE 68544

Serendipity
PO Box 2450
Amarillo, TX 79101

Current, Inc
Colorado Springs, Colorado 80941

Propinquity
8915 Santa Monica Blvd
Los Angeles, CA 90069

Spencer Gifts. Inc.
1050 Black Horse Pike
Atlantic City, NJ 08411

Brentano's
586 Fifth Ave
New York, NY 10036

Hampacher-Schlemmor
39-25 Skillman Ave
Long Island City, NY 11104

Amsterdam Company
Wallins Corner Road
Amsterdam, NY 12010

Taylor Gifts
Conestoga Rd & Lancaster Ave
Strafford Wayne, PA 19087

Unity Buying Service
5675 Jimmy Carter Blvd
Norcross, GA 30091

Abbey Press
St Mainrad, Indiana 47577

Miles Kimball Co
41 W Eighth Ave
Oshkosh, WI 54901